FLORA OF TROPICAL EAST AFRICA

CONNARACEAE

T0132821

J. H. HEMSLEY

Erect or scandent shrubs, small trees or lianes. Leaves alternate, uni-foliolate or imparipinnate; stipules and stipellae absent. Inflorescence terminal or axillary, paniculate or racemose, often fascicled in axils of older leaves or on older branchlets. Flowers hermaphrodite, rarely dioecious,* actinomorphic, usually pentacyclic, androecium and gynoecium often dimorphic or heteromorphic. Sepals 5, imbricate or valvate, usually free. Petals 5, free or lightly connate near base. Stamens 10, 5 episepalous longer than 5 epipetalous, the latter sometimes reduced and staminodal; filaments free or connate into a short tube at base; anthers dorsifixed, usually with longitudinal introrse dehiscence. Carpels 5 or solitary, rarely 3, free; ovules normally 2, basal or attached to ventral suture, collateral, anatropous or orthotropous, erect. All carpels apparently fertile but usually only 1–3 mature. Fruit usually follicular, dehiscence by ventral suture, sometimes indehiscent or dehiscence irregular. Seed solitary, with ± fleshy aril covering part or all of the seed; aril free or sometimes fused with testa to form a compound fleshy pseudo-aril. Endosperm abundant, rudimentary or absent; cotyledons plano-convex, fleshy, or much flattened and small; radicle apical or laterally displaced.

Leaves unifoliolate :
  Lower surface of leaflet with dense small dark glandular
    dots . . . . **3. Vismianthus**
  Lower surface of leaflet without glandular dots :
    Petiole short, ± 1 cm. in length . . . **8. Ellipanthus**
    Petiole long, up to half length of leaf-lamina . **2. Burttia**
Leaves compound :
  Leaves trifoliolate . . . . . . **4. Agelaea**
  Leaves imparipinnate with 2–15 pairs of leaflets :
    Carpels 5, sepals and petals without glandular dots :
      Follicle velutinous and with urticating hairs, apex
        narrowed to a long rostrum . . . **1. Cnestis**
      Follicle glabrescent, apex acute to rounded :
        Flowering precociously or with new crop of
          leaves; inflorescence axillary, racemose,
          few-flowered . . . . . **6. Eyrsocarpus**
        Flowering not precocious; inflorescence ter-
          minal, sub-terminal or axillary, paniculate,
          many-flowered :
        Follicle dehiscent by simple ventral suture,
          pericarp surface smooth . . . **7. Jaundea**
        Follicle indehiscent or splitting by irregular
          basal fissures, surface with fine longi-
          tudinal striation . . . . **5. Santaloïdes**

---

* Not, as far as is known, in African species.

Carpel solitary, sepals and petals with conspicuous
    dark glandular dots and streaks   .   .   .   9. **Connarus**

NOTE. The relationships of stamen and style lengths present much interest in the
E. African material. In nearly all species examined there is present a well-marked
dimorphy, i.e., the flowers fall into two groups, those with long stamens and short
styles, and those with short stamens and long styles. Examination of material and
literature gives little evidence of loss of function in either form of flower, but it is
difficult to avoid the conclusion that there is a well-marked tendency towards dio-
ecism within the group. It would be interesting to know whether the dimorphic
habit is a fixed one, e.g., does any one individual always bear the same kind of
flower throughout its life, and do plants with the short, ill-developed style set fruit
on the same scale as those with the long well-developed style ? Such problems await
the attention of the field observer.

1. CNESTIS

Juss., Gen. Pl. 374 (1789) ; Schellenb. in E.P. IV. 127 : 29 (1938)

Shrubs, scandent or otherwise, small trees or lianes. Leaves imparipinnate ; leaflets opposite or subopposite, entire. Inflorescence terminal or
axillary, paniculate or racemose ; racemes often clustered in old leaf-bases
on older branches. Flowers pentamerous, androecium and gynoecium
dimorphic. Sepals free, valvate or slightly imbricate. Petals shorter than,
as long as, or longer than sepals. Stamens 10 ; filaments free or connate
at base into short tube. Carpels densely hairy, stigma expanded, rounded
or capitate ; ovules inserted basally or on ventral suture near base. Fruit
a pyriform follicle, the apex usually produced into a curved or twisted
rostrum ; dehiscence by ventral suture ; pericarp densely velutinous and
hairy externally, hairy internally, the hairs often urticating. Seed ovoid ;
testa smooth and shining, aril free, basal, oblique, thin, not conspicuous ;
cotyledons flattened ; endosperm present, fleshy.

An old world genus, mainly tropical, the majority of species being found in tropical
and sub-tropical Africa.

Leaves large, to 35 cm. long ; leaflets up to 15 pairs,
    oblong-elliptic to narrowly oblong   .   .   .   1. *C. ugandensis*
Leaves smaller, to 21 cm. long, leaflets up to 8 pairs,
    ovate to ovate-elliptic :
    Leaflets up to 8 pairs, usually 4–6, lower surface with
        few scattered pale hairs   .   .   .   .   .   2. *C. confertiflora*
    Leaflets 3 pairs, lower surface ferruginous tomentose .   3. *C. calocarpa*

1. **C. ugandensis** *Schellenb.* in E.P. IV. 127 : 44 (1938) ; I.T.U., ed. 2,
100 (1952). Type : Uganda, Mengo District, Mabira Forest, Mulange,
Dummer 5422 (K, holo. !)

Small tree to 7 m. high, young branches densely pubescent, older branches
becoming glabrous. Leaf-rhachis to 35 cm. long with dense greyish brown
indumentum. Leaflets to 15 pairs, oblong-elliptic to narrowly oblong, 3·5–
14 cm. long, 1·5–3 cm. wide, apex rounded to sub-acuminate, base rounded,
oblique, chartaceous ; upper surface ± glabrous, lower surface with dense
cinerous pubescence ; lateral nerves 7–9 pairs, arcuate, ascending. Inflorescences axillary, densely fascicled on older branchlets, racemose ;
rhachis to 13 cm. long ; rhachis, bracts, pedicels and sepals densely pubescent. Sepals narrowly elliptic, 6·5 mm. long, 2 mm. wide. Petals white or
creamish, narrowly elliptic, 7·5 mm. long, 2·5 mm. wide, glabrous. Stamens
10 ; long-stamened flowers, 5 episepalous stamens to 6 mm. long, 5 epipetalous to 4 mm. long ; short-stamened flowers, 5 episepalous stamens to
2 mm. long, 5 epipetalous to 1·3 mm. long, filaments free, flattened, glabrous.

Fig. 1. *CNESTIS CONFERTIFLORA*—**1,** leafy branch with fruits, × ⅔ ; **2,** leaflet to show variation in size and shape, × ⅔ ; **3,** inflorescence, × 1½ ; **4,** long-styled flower, × 4½ ; **5,** same flower with some sepals and petals removed, × 9 ; **6,** long-stamened flower, × 4½ ; **7,** same flower with some sepals and petals removed, × 9 ; **8,** fruit, l.s., showing insertion of seed, × 1 ; **9,** seed, × 1½. 1, 4, 5 and 9, from *Kirk* s.n., 4/70 ; 2, from *Greenway* 2745 ; 3, from *Kirk* s.n., 5/70 ; 6 & 7, from *Mason* s.n. recd. Aug. 1912 ; 8, from *Holtz* 157.

Ovary ovoid, ± 0·5 mm. long ; styles of long-stamened flowers to 1 mm. long, strongly recurved ; those of short-stamened flowers to 4 mm. long, puberulous. Ovules basally affixed. Fruit a scarlet pyriform follicle to 2·5 cm. long, 1·2 cm. diam. ; rostrum ± 1 cm. long; surface longitudinally rugose and densely beset with short stiff bristles. Mature seeds not seen.

UGANDA. Bunyoro District : Budongo forest, Dec. 1934 (fl.), *Eggeling* 1468 !
DISTR. U2, 4 ; northwards into the A.-E. Sudan
HAB. Understorey tree in rain-forest, 900–1300 m.

NOTE. The type gathering of this species seems to be of unusual vigour, with large inflorescences and flower buds. Future collections may give smaller measurements for flower parts than stated here ; present material is very scanty and further gatherings, especially of mature fruits and seeds, are needed in order to give a more complete account of the plant.

2. **C. confertiflora** *Gilg* in P.O.A. C : 193 (1895) ; E.P. IV. 127 : 45 (1938) ; T.T.C.L. 168 (1949). Type : Tanganyika, Uzaramo District, Kisserawe, *Stuhlmann* 6262 (B, holo. †)

Scandent shrub or small tree, height to 7 m., young branches pubescent, older branches glabrous. Leaf-rhachis 6–21 cm. long ; pubescent or puberulous. Leaflets 3–8 (usually 4–6) pairs, ovate to ovate-oblong, terminal leaflet tending to obovate, (1·5–) 3–6 (–12) cm. long, (1·4–) 1·7–2·6 (–6) cm. wide, apex broadly acute to acuminate, base obliquely subcordate with the margin slightly overlapping or running parallel with rhachis, chartaceous to subcoriaceous ; upper surface ± glabrous, lower surface with few scattered hairs, denser on midrib. Inflorescences clustered on short lateral shoots, racemose ; rhachis 1·5–4·5 cm. long, densely hairy. Sepals narrowly lanceolate, 3·5 mm. long, 1 mm. wide, outer surface pilose, inner glabrous. Petals white, narrowly elliptic, to 8 mm. long, 2 mm. wide, glabrous. Stamens 10 ; long-stamened flowers, 5 episepalous stamens to 4 mm. long, 5 epipetalous to 3 mm. long ; short-stamened flowers, 5 episepalous stamens to 1·7 mm. long, 5 epipetalous to 1·2 mm. long, filaments free, flattened near base, glabrous. Ovary subglobose, 0·5 mm. long ; styles of long-stamened flowers to 0·5 mm. long, recurved, styles of short-stamened flowers to 3·5 mm. long, pilose. Ovules inserted near base of ventral suture. Fruit a pyriform follicle (see Fig. 1/1 and 1/8), to 1·5 cm. long, 1 cm. diam., rostrum to 1·8 cm. long ; pericarp bright red, surface longitudinally furrowed, with dense minute papillate protuberances and caducous urticating bristles ; internal wall of follicle densely hairy. Seed shining black, ovoid and somewhat flattened, 1·3 cm. long, 9 mm. diam. ; aril 4 mm. long, oblique, slightly lobed. Fig. 1, p. 3.

TANGANYIKA. Morogoro District : Kimboza forest, Apr. 1954 (fl.), *Padwa* 331 ! ; Uzaramo District : Dar es Salaam, May 1870 (fl. & young fr.), *Kirk* !
ZANZIBAR. Pemba Island, Tasini, 17 Dec. 1930 (fl.), *Greenway* 2745 !
DISTR. T6 ; P ; not known elsewhere
HAB. Coastal evergreen bushland and riverine forest, 0–300 m.
SYN. *C. riparia* Gilg in E.J. 23 : 217 (1896). Type : Tanganyika, Uluguru foothills, Luhangulo in riverine forest of Ruvu, *Stuhlmann* 8942 (B, holo. †)
     C. confertiflora Gilg forma *macrophylla* Schellenb. in E.P. IV. 127 : 45 (1938). Type : Tanganyika, Uzaramo District, Pugu Hills, *Holtz* 2059 (B, holo. †)

NOTE. *C. riparia* and *C. confertiflora* forma *macrophylla* are based on single specimens and depend upon minor differences of leaflet shape and size for diagnostic character. It is felt that both can be encompassed within the present species and are included in synonymy above. It is difficult to separate material determined as *C. lescrauwaetii* De Wild., a Lower Congo and Angolan species, from *C. confertiflora* Gilg. The former has somewhat larger fruits (follicle 2 cm. long, rostrum 2·5 cm. long), but in the absence of material linking these two widely separated and low altitudinal areas, it seems inadvisable to include *C. lescrauwaetii* De Wild. as a synonym at the present time.

3. C. calocarpa *Gilg* in P.O.A. C : 192 (1895) ; E.P. IV. 127 : 46 (1938) ; T.T.C.L. 168 (1949). Type : Tanganyika, Uzaramo District, Mgambo, *Stuhlmann* 6388 (B, holo. †)

Young branches tomentose. Leaf-rhachis 10–13 cm. long, leaflets 3 pairs, ovate-oblong, 2–8 cm. long, 1·5–3·7 cm. wide, apex shortly acuminate, base obliquely cordate, membranous or subcoriaceous ; glabrous and shining above, densely ferruginous tomentose beneath, lateral nerves 7–9 pairs. Inflorescences racemose, to 6 cm. long, fascicled in axils of fallen leaves, rhachis ferruginous tomentose. Sepals 3 mm. long, tomentose externally, glabrous internally. Petals not present. 3–5 carpels mature; follicle ± 1·5 cm. long, 1 cm. diam., apex tapered into a long beak-like process 1·2 cm. long ; pericarp reddish, tomentose, beset with brownish caducous urticating hairs. Seed subglabrous, ± 1 cm. long ; testa shining black, aril 3 mm. long, basal, unilateral.

TANGANYIKA. Uzaramo District : Mgambo [about 25 km. S. of Bagamoyo], *Stuhlmann* 6388 & Yegea, *Stuhlmann* 8623
DISTR. T6 ; not known elsewhere
HAB. Not known

NOTE. No authentic material of this species has been seen and there is doubt whether it is really distinct. The above account has been taken from Gilg's original description and from Schellenberg in the Pflanzenreich, the latter author having seen both specimens cited. For the present the species is kept up, the small number of leaflets and the dense ferruginous indumentum of the leaflet lower-surface being the main points of divergence from *C. confertiflora* Gilg which has the hairs very sparsely scattered and pale in colour. Further gatherings from the type-area, to show particularly the indumentum and leaflet-number variation, are necessary to decide the true identity of this plant.

2. **BURTTIA**

Bak. f. & Exell in J.B. 69 : 249 (1931)

Shrub or small tree with unifoliolate, long-petioled leaves. Inflorescence racemose, few-flowered, axillary. Flowers with dimorphic androecium and gynoecium. Sepals 5, imbricate in bud, very shortly connate at base, persisting in fruit. Petals 5, free, subequal, larger than sepals. Stamens 10, five episepalous longer than five epipetalous ; filaments connate for short distance at base. Carpel solitary ; ovary densely hairy ; style simple with broadened papillose stigma. Ovules 1–3, usually 2, inserted towards apex of ventral suture. Fruit a subcylindrical follicle, densely pubescent. Dehiscence by ventral suture. Seed narrowly ovoid ; aril lobate, spreading laterally from sides of hilum ; embryo straight ; cotyledons flattened, embedded in copious endosperm.

A monotypic genus restricted to Tanganyika and Northern Rhodesia, unique in the family in the combination of unifoliolate leaves and solitary carpel with the possession of abundant endosperm in the seed.

B. prunoïdes *Bak. f. & Exell* in J.B. 69 : 249 with fig. (1931) ; Schellenb. in E.P. IV. 127 : 96, fig. 14 (1938) ; T.T.C.L. 167 (1949). Type : Tanganyika, Singida District, Itigi–Saranda–Kasikasi area, *B. D. Burtt* 532 (BM, holo. !, EA, iso. !)

Deciduous bushy shrub or small tree to 6 m. high. Young branches pubescent, later becoming glabrous ; shoots of unlimited growth with little scarring, short shoots with conspicuous annular scars. Leaves in terminal rosettes on long and short shoots ; petiole 2·5–5 cm. long, densely rusty pilose when young, later glabrescent ; lamina ovate, broadly elliptic

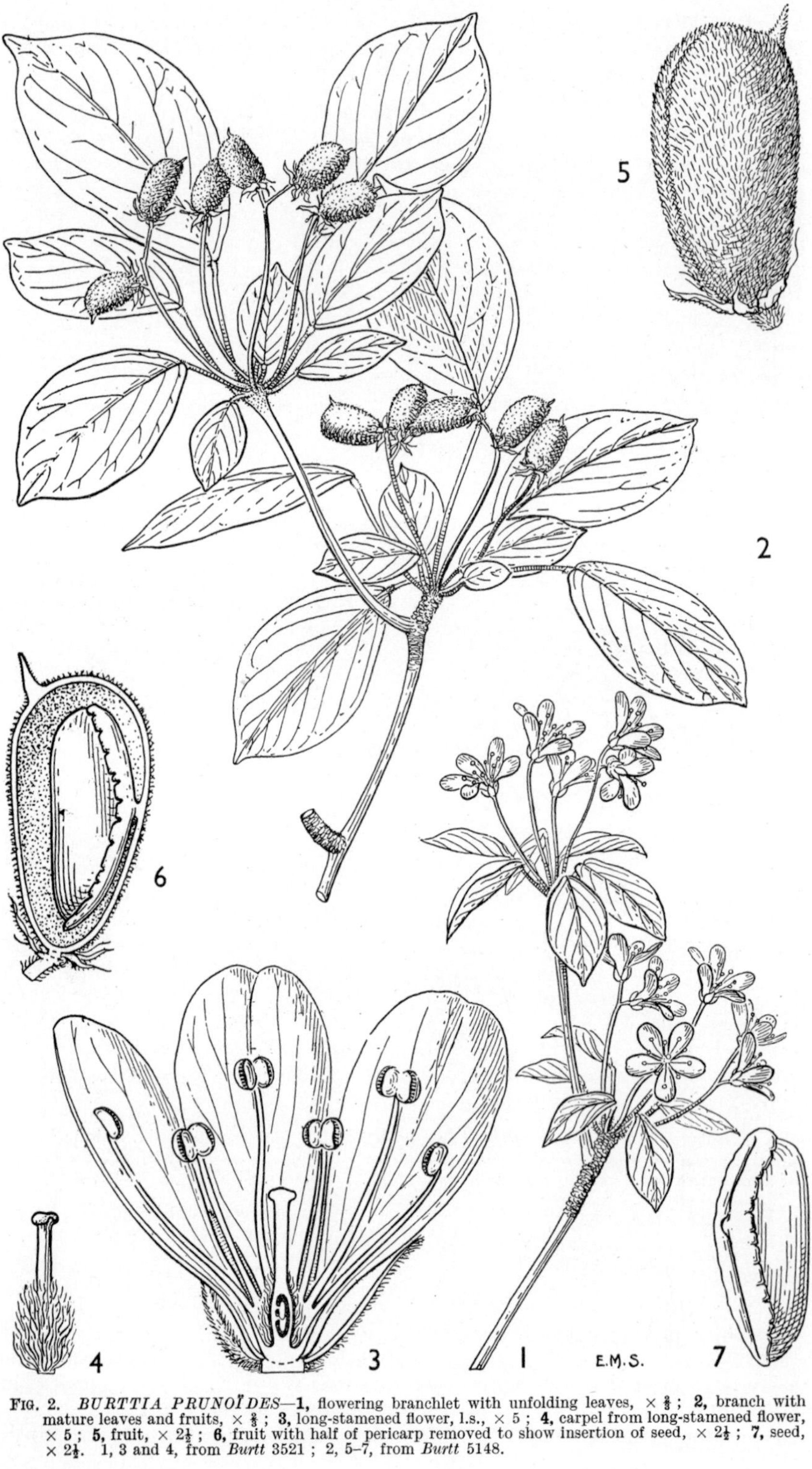

FIG. 2. *BURTTIA PRUNOĬDES*—**1,** flowering branchlet with unfolding leaves, × ⅔ ; **2,** branch with
mature leaves and fruits, × ⅔ ; **3,** long-stamened flower, l.s., × 5 ; **4,** carpel from long-stamened flower,
× 5 ; **5,** fruit, × 2½ ; **6,** fruit with half of pericarp removed to show insertion of seed, × 2½ ; **7,** seed,
× 2½. 1, 3 and 4, from *Burtt* 3521 ; 2, 5–7, from *Burtt* 5148.

to suborbicular, 5–8 cm. long, 3–6 cm. wide, apex obtuse and shortly acuminate, base rounded to subcordate ; upper surface puberulous when young,
later glabrous, lower surface pubescent when young, hairy on midrib and
main nerves only in older leaves ; lateral nerves 6–9 pairs, ascending.
Racemes congested in axils of young leaf rosette ; rhachis slender, to 5 cm.
long, rusty hairy. Sepals oblong to oblong-elliptic, 4–6 mm. long, 2–3 mm.
wide, rusty pubescent. Petals white, ± obovate, 9–13 mm. long, 5–7 mm.
wide, glabrous. Long-stamened flowers with episepalous stamens to 10 mm.
long and epipetalous to 7 mm. long ; short-stamened flowers with episepalous stamens to 5 mm. long and epipetalous to 3·5 mm. long; filaments
glabrous. Ovary to 2 mm. long, 1·5 mm. wide, densely pilose ; style of long-
stamened flowers to 3 mm. long, that of short-stamened flowers to 6 mm.
long, glabrous ; stigma papillose. Fruit to 1·3 cm. long, 9 mm. diam., apex
with a short rostrum, calyx persisting at base, sericeous. Seed to 1·2 cm.
long, 6 mm. diam. ; aril bright red at maturity. Fig. 2.

TANGANYIKA. Shinyanga District : Tinde Hills, Usanda Hill, Feb. 1935 (fr.), *B. D.
    Burtt* 5148 ! ; Dodoma District : road to Kondoa, 29 Mar. 1928 (fr.), *B. D. Burtt*
    1800 ! ; Kondoa District : Sambala, Wamkuna Hills, 24 Mar. 1929 (fr.), *B. D.
    Burtt* 1978 !
DISTR. T1, 5 ; Northern Rhodesia
HAB. Deciduous thicket, 800–1500 m.

NOTE. When flowering this species bears a close resemblance to the genus *Prunus,*
    in particular the cultivated cherry ; the abundant white flowers open with the young
    leaves in the December rains and the plant fruits in February and March. It appears
    to be common in the Central Province of Tanganyika and reappears again in Northern
    Rhodesia, no material seems to have been obtained from the Western Province of
    Tanganyika where the species almost certainly occurs on the Tanganyika side of the
    Kalambo falls and river.

3. **VISMIANTHUS**

Mildbr. in N.B.G.B. 12 : 706 (1935)

Shrub with unifoliolate long-petioled leaves, terminally arranged on
long and short shoots. Inflorescence axillary, few flowered. Sepals and
petals 5, with dark glandular dots and streaks. Stamens 10 ; filaments
glabrous, free ; anthers ovoid, ± apiculate. Carpel solitary, densely pilose ;
stigma subcapitate. Ovules affixed to ventral suture in a median position.
Fruit a flattened ± ellipsoid follicle, tapering at each end ; pericarp with
thin glabrous outer wall, and thicker cartilaginous inner wall. Seed ovoid,
hilum lateral towards base, series of small fimbriate processes present at
base (see note, page 9 and Fig. 3/12, 3/12a).

A monotypic genus known only from the Southern Province, Tanganyika.

V. punctatus *Mildbr.* in N.B.G.B. 12 : 706 (1935) ; E.P. IV. 127 : 98,
fig. 15 (1938) ; T.T.C.L. 169 (1949). Type : Tanganyika, Lindi District,
Mlinguru, about 20 km. S. of Lindi, *Schlieben* 5757 (B, holo. †, BM, iso. !)

Much branched shrub to 5 m. high, young branches pubescent, later
becoming glabrous with pale brown smooth bark and inconspicuous trans-
verse ringing. Petiole slender, about half length of leaf blade, puberulous ;
lamina broadly ovate, to 8 cm. long, 5 cm. wide, apex long and narrowly
acuminate, base slightly cordate, chartaceous ; upper surface sparsely
puberulous when young, later glabrescent ; lower surface with scattered
bifurcate hairs, denser on midrib and main lateral nerves, thickly beset with
small dark glandular dots ; lateral nerves 4–6, arcuate, ascending. Inflorescence raceme-like, with about 5 flowers ; rhachis 1·5–5 cm. long,
puberulous. Sepals green with dark red glandular dots and streaks, nar-

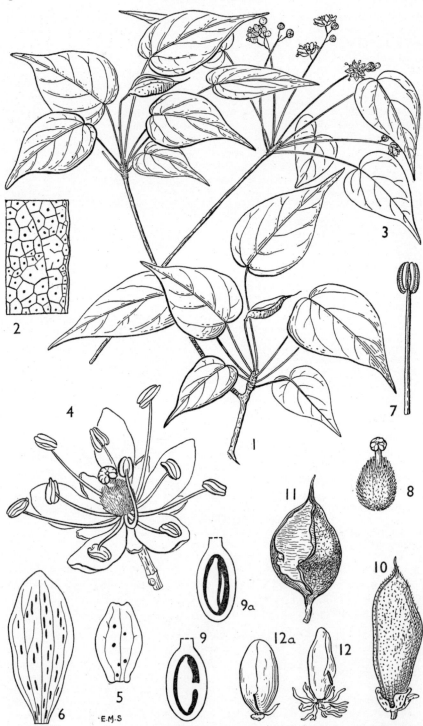

FIG. 3. *VISMIANTHUS PUNCTATUS*—**1**, leafy branch, × ⅔ ; **2**, lower surface of leaf to show gland dots, × 4 ; **3**, flowering branch, × ⅔ ; **4**, flower, × 8 ; **5**, sepal, × 8 ; **6**, petal, × 8 ; **7**, episepalous stamen, × 8 ; **8**, carpel, × 8 ; **9** and **9a**, diagrammatic l.s. ovary to show position and insertion of ovules, × 16 ; **10**, immature fruit, × 2 ; **11**, fruit showing premature dehiscence, × 2 ; **12** and **12a**, immature seeds, × 4.   1, 2, 11, 12 and 12a, from *Semsei* 647 ; 3–9a, from *Eggeling* 6402 ; 10, from *Gillman* 1315.

rowly elliptic and deeply concave, to 3·5 mm. long, 1·5 mm. wide, ± glabrous. Petals white with lines of dark red glandular dots and streaks, obovate to elliptic-oblong, to 5 mm. long, 2·3 mm. wide, glabrous. Five episepalous stamens to 3·5 mm. long, 5 epipetalous to 2·5 mm. long. Ovary ovoid, to 1 mm. long, densely covered with long pale brown hairs, style ± 2 mm. long. Fruit (not fully mature) to 1·7 cm. long, 8 mm. diam., apex narrowed to a slender rostrum ; dehiscence probably by ventral suture ; outer layer of pericarp thin and brittle when dry, puberulous when young, later glabrescent, with prominent raised nerves and scattered dark raised gland-dots. Seed solitary, with thickened attachment to ventral placenta. Fig. 3.

TANGANYIKA. Lindi District : Rondo Plateau, Mchinjiri, Nov. 1951 (fl.), *Eggeling* 6402 ! & Feb. 1952 (fr.), *Semsei* 647 ! ; Newala District : Kitangari, *Gillman* 1315 !
DISTR. T8 ; not known elsewhere
HAB. Coastal evergreen bushland and secondary forest on margins of native cultivation, 800 m.

NOTE. This monotypic genus, known only from the Southern Province, Tanganyika, appears to be frequent on the Rondo and Makonde plateaux. Ripe fruits and mature seeds have not yet been obtained but the immature seed shows the presence of a very interesting basal structure. This takes the form of a series of branched processes, arising apparently from the testa, which are suggestive of a suspensor mechanism comparable with that found in *Schellenbergia sterculiifolia* (Prain) Parkinson, a monotypic Malayan genus. The basal processes although appearing to be quite separate from the hilum seem to be of arillar derivation. Observations on the method of fruit dehiscence and collections to include mature fruit and seeds are required to supplement the very sparse herbarium material available of this plant.

4. AGELAEA

[Soland. ex] Planch. in Linnaea 23 : 437 (1850) ; Schellenb. in E.P. IV. 127 : 65 (1938)

Climbing shrubs, small trees or lianes ; young branches, inflorescences, etc., usually with stiff fascicled hairs. Leaves trifoliolate, leaflets usually entire, lateral leaflets oblique. Inflorescence a large terminal or axillary many-flowered panicle. Flowers pentamerous, androecium and gynoecium heteromorphic. Sepals imbricate, with dense brown indumentum, usually fringed with glandular hairs. Petals longer than sepals, often slightly connate at base, glabrous. Filaments connate at base, terete, glabrous. Carpels pilose ; stigma expanded and sometimes flattened ; ovules basally inserted. Fruit an obovoid or pyriform follicle, base constricted ; dehiscence by ventral suture, pericarp peeling and curling back to expose seed ; pericarp velutinous. Seed ± ovoid, testa dark and shining ; aril basal, obliquely cupuliform, somewhat fleshy, cotyledons plano-convex, radicle apical, endosperm nil.

This genus is mainly tropical African in distribution with extensions southwards to Portuguese East Africa, the Rhodesias and Angola, and northwards into the A.-E. Sudan ; four species are recorded from the Mascarene Islands.

NOTE. In addition to the long- and short-stamened flowers, this genus shows a third type of stamen-style relationship. *A. heterophylla* Gilg and *A. ugandensis* Schellenb., possess flowers in which the long episepalous stamens are coupled with a long style, the epipetalous stamens however, remain small and resemble those of typical short-stamened flowers. Reference to Fig. 4/7, 4/8 and 4/9 will show the three sorts of flower found in *A. ugandensis*.

Lower surface of leaflet with uniformly scattered small
  fascicled hairs ; nerves, midrib and leaf-rhachis
  with dense short indumentum . . . . 1. *A. heterophylla*

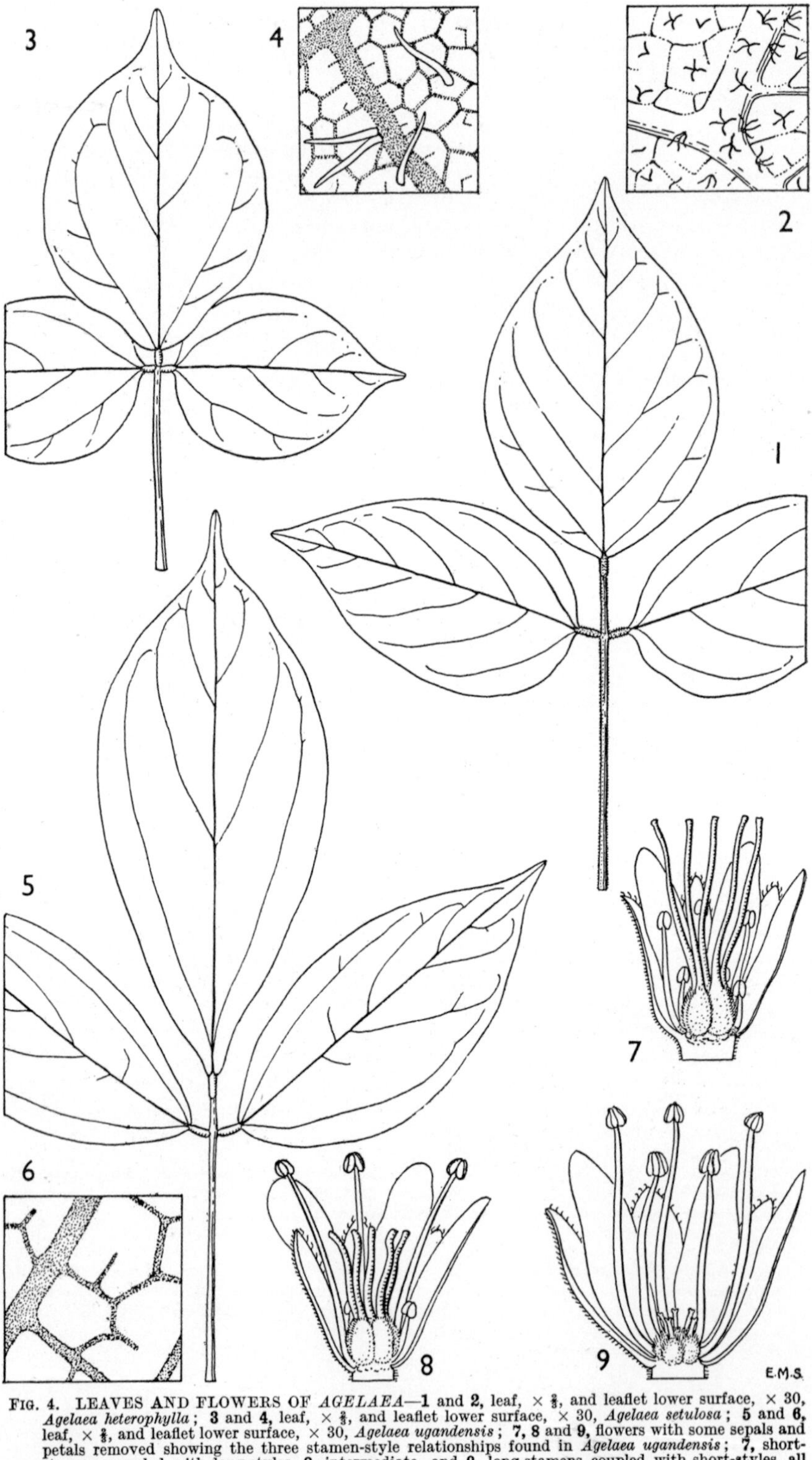

FIG. 4. LEAVES AND FLOWERS OF *AGELAEA*—**1** and **2**, leaf, × ⅔, and leaflet lower surface, × 30, *Agelaea heterophylla*; **3** and **4**, leaf, × ⅔, and leaflet lower surface, × 30, *Agelaea setulosa*; **5** and **6**, leaf, × ⅔, and leaflet lower surface, × 30, *Agelaea ugandensis*; **7, 8** and **9**, flowers with some sepals and petals removed showing the three stamen-style relationships found in *Agelaea ugandensis*; **7**, short-stamens coupled with long-styles, **8**, intermediate, and **9**, long-stamens coupled with short-styles, all × 6. 1 and 2, from *Greenway* 1046; 3 and 4, from *Scheffler* 247; 5 and 6, from *Wood* 174; 7, from *Dale* 3126; 8, from *Eggeling* 1523; 9, from *Purseglove* 3523.

Lower surface of leaflet glabrous or with few scattered
    simple, paired or fascicled hairs ; nerves, midrib
    and leaf-rhachis with scattered longer hairs, becom-
    ing subglabrous with age :
Terminal leaflet ovate-elliptic to obovate ; basal pair
    of nerves strongly developed, angle of divergence
    very acute .   .   .   .   .   .   2. *A. ugandensis*
Terminal leaflet broadly ovate to suborbicular ; basal
    pair of nerves usually giving rise to well-marked
    secondary branches at ± 2 cm. from base, angle
    of divergence less acute .   .   .   .   3. *A. setulosa*

1. **A. heterophylla** *Gilg* in N.B.G.B. 1 : 66 (1895) ; Schellenb. in E.P. IV.
127 : 75, fig. 11 (1938) ; T.T.C.L. 167 (1949). Type : Tanganyika, Morogoro, *Stuhlmann* (B, holo. †)

Shrub, small tree or liane to 30 m. Young branches with dense brown
indumentum, consisting of short stiff fascicled hairs projecting slightly above
pale-coloured weaker fascicled hairs, persisting as greyish arachnoid
covering but old stems becoming subglabrous and purple-brown in colour.
Leaf-rhachis 2·5–18 cm. long, indumentum as above. Terminal leaflet
obovate to broadly elliptic, 6–18 cm. long, 3·5–14 cm. wide, apex shortly
acuminate, base rounded to cuneate, rigidly chartaceous ; midrib chan-
nelled above with fascicled hairs along length ; lower surface of leaflet with
scattered fascicled hairs, usually 4-armed, dense on midrib and nerves ;
lateral nerves to 6 pairs, arcuate ascending (see Fig. 4/1 and 4/2). Inflorescence terminal, to 30 cm. long ; rhachis, bracts, bracteoles and pedicels
densely pubescent as on young stems. Flowers fragrant. Sepals lanceolate
to oblong-lanceolate, ± 5 mm. long, 1–1·5 mm. wide, fringed with reddish
glandular hairs. Petals white, elliptic to oblong, to 5·5 mm. long, 2 mm.
wide. Stamens 10 ; long-stamened flowers, 5 episepalous stamens to 5 mm.
long, 5 epipetalous to 3 mm. long ; short-stamened flowers, 5 episepalous
stamens to 3 mm. long, 5 epipetalous to 1·5 mm. long (see note, page 9) ;
filament-tube to 1·5 mm. long ; anthers apiculate. Ovary ovoid, ± 1 mm.
long, densely pilose ; styles of short-stamened flowers to 5 mm. long, long-
stamened flowers to 1·5 mm. long, puberulous. Fruit a ± obovoid follicle,
to 2 cm. long, 1·4 cm. diam. ; pericarp red, with dense velvety pubescence.
Seed black, to 1·4 cm. long, 8 mm. diam. ; aril white, to 6 mm. long, lobulate
and somewhat fleshy.

UGANDA. Bunyoro District : Bugoma Forest, Feb. 1943 (fl.), *Purseglove* 1247 !
KENYA. E. side of Mt. Kenya, 29 July 1913 (fl. & young fr.), *Battiscombe* 691 ! ; Teita
    Hills, Ngerenyi, below Verbi's House, 7 Feb. 1953 (fr.), *Bally* 8783 !
TANGANYIKA. Lushoto District : Western Usambara Mts., Bumbuli–Mazumbai road,
    8 May 1953 (fl.), *Drummond & Hemsley* 2437 !, Kwamshemshi–Sakare road, 4 July
    1953 (fr.), *Drummond & Hemsley* 3153 ! ; Morogoro District, probably above Moro-
    goro, 21 Nov. 1932 (fr.), *Wallace* 463 !
DISTR. U2, 4 ; K4, 7 ; T1–3, 6, 7 ; southwards to Portuguese East Africa and the
    Rhodesias
HAB. Lowland and upland rain-forests; this species seems to thrive in the regeneration
    of partially cut-out forest and fruits abundantly under these conditions, 900–2100 m.

SYN. [*A. usambarensis* [Gilg ex] Engl., P.O.A. C : 86 (1895), *in obs.*, *nomen nudum*]
    A. obliqua (Beauv.) Baill. var. *usambarensis* Gilg in P.O.A. C : 192 (1895).
    Type : Tanganyika, E. Usambara Mts., Nderema, 23 Feb. 1893, *Holst* 2234
    (K, iso. !)
    [*A. obliqua* sensu Battiscombe in T.S.K. 85 (1926) and Dale in T.S.K. 113 (1936) ;
    T.T.C.L., part 1, 21 (1940), *non* (Beauv.) Baill.]

VARIATION. A large range of leaflet-size is to be found in this species. Small bushy
    saplings which occur in quantity in the shrub-layer of most of the E. Tanganyika
    upland rain-forests have the largest leaflets whilst the smaller and rigidly chartaceous

leaflets usually come from the top of the liane, either from the forest-canopy or on the forest-margins under conditions of higher light intensity. There is also a range of filament connation length, the longest staminal-tube being found in Uganda specimens and a progressive decrease is found southwards through Kenya to Tanganyika.

2. **A. ugandensis** Schellenb. in E.J. 58 : 219 (1923) ; [Schellenb. in V.E. 3 (1) : 322 (1915), in obs., nom. nud.] ; E.P. IV. 127 : 82 (1938) ; T.T.C.L. 167 (1949). Types : Uganda, no precise locality, *Scott Elliot* 7397 (K, syn. !) ; Masaka District, Buddu, *Dawe* 271 (K, syn. !) & Lake Victoria, Sese Islands, *Stuhlmann* 1227 (B, syn. †) and Tanganyika, Bukoba, *Stuhlmann* 1573 (B, syn. †)

Shrub, scandent or otherwise, small tree or liane to 30 m. high. Young branchlets with brown pubescence ; hairs of two kinds, stiff smooth brown hairs, fascicled in clusters of 4, projecting above smaller simple or fascicled hairs ; old stems subglabrous and dull purplish in colour. Leaf-rhachis 2–12 cm. long, pilose when young, becoming subglabrous ; terminal leaflet ovate-elliptic to obovate, 4–12 (–21) cm. long, 2–5·5 (–9) cm. wide, apex shortly acuminate, base rounded to cuneate ; lateral leaflets ovate to broadly elliptic, all rigidly chartaceous ; upper surface glabrous or with few hairs along midrib, lower surface subglabrous or with scattered simple or fascicled hairs on midrib and main nerves ; lateral nerves 3–5 pairs, arcuate ascending, basal pair very prominent (see Fig. 4/5 & 4/6, p. 10). Inflorescence terminal, to 30 cm. long ; rhachis, branchlets, bracts and pedicels with dense brown indumentum as on young stems. Flowers fragrant. Sepals broadly lanceolate, to 4·5 mm. long, 2 mm. wide, fringed with reddish glandular hairs. Petals white, elliptic-oblong, to 6 mm. long and 2 mm. wide. Stamens 10 ; long-stamened flowers, 5 episepalous stamens to 5·5 mm. long, 5 epipetalous to 3·5 mm. long ; short-stamened flowers, 5 episepalous stamens to 3·5 mm. long, 5 epipetalous to 1·5 mm. long (but see note on p. 9 and Fig. 4/7, 4/8 & 4/9, p. 10), filament-tube short, anthers ± apiculate. Ovary ovoid, ± 1 mm. long, pilose ; styles of short-stamened flowers to 5·5 mm. long, long-stamened flowers to 1·5 mm. long, puberulous. Fruit a red obliquely obovoid follicle, to 1·5 cm. long, 9 mm. diam., densely pubescent with short stiff fascicled hairs. Seed shining black, ovoid-oblong, 1·2 cm. long, 7 mm. diam. ; aril white, to 4 mm. long, lobulate.

UGANDA. Kigezi District : Buambara, Nov. 1950 (fl.), *Purseglove* 3523! ; Mengo District : Entebbe, Kitubilu Forest, Apr. 1923 (fr.), *Maitland* 716!
KENYA. N. Kavirondo District : Kakamega Forest, Apr. 1933 (fl.), *Dale* 3126 in C.M. 14589! ; S. Kavirondo District : Kuja River, 15 July 1946 (fl.), *Glasgow* 46/33!
TANGANYIKA. Bukoba District : Bushasha, 1935, *Gillman* 313! ; Mwanza District : Ukerewe Island, 15 Dec. 1926 (fl.), *Conrads* 520!
DISTR. U2, 3, 4 ; K5 ; T1 ; and southern A.-E. Sudan
HAB. Primarily a liane of lowland and upland rain-forest but often found in relict and in regeneration complexes of these forests, also in riverine forest, 1100–1700 m.

SYN. [*A. obliqua* sensu auct. plur., ex parte, non (Beauv.) Baill.]
     [*A. nitida* sensu auct. in Check List Uganda Trees and Shrubs 39 (1935), based on *Bagshawe* 609, non Planch.]

NOTE. This species shows a third stamen-style relationship, *Eggeling* 1523 from Budongo Forest, Bunyoro District, *Wood* 324 from Kimaka Hill, about 6·5 km. N. of Jinja and *Gillman* 313 cited above, have the long episepalous stamens coupled with a medium-long style (Fig. 4/8, p. 10). Flowers from long- and short-stamened specimens (Fig. 4/7 & 4/9, p. 10) are given for comparison. It is interesting to note that these gatherings are from plants said to be bushes or shrubs and there may be some correlation between form of flower and the stage in development of the individual plant.

3. **A. setulosa** Schellenb. in E.J. 58 : 211 (1923) ; E.P. IV. 127 : 81 (1938) ; T.T.C.L. 167 (1949). Type : Tanganyika, Pangani District, Makinjumbi on the Pangani River, *Scheffler* 247 (B, holo. †, K, iso. !)

Scandent shrub or liane to 20 m. high. Young branches pubescent, hairs ± dense, setulose, brown and usually paired, older branches glabrescent, dark reddish brown. Leaf-rhachis 3–12 cm. long, indumentum same as on young branches, soon becoming glabrescent ; petiolules with simple or paired setulose hairs ; terminal leaflet broadly ovate to suborbicular, 4–12 cm. long, 3–11·5 cm. wide, apex abruptly acuminate, base rounded, rigidly chartaceous ; upper surface glabrous, midrib with a few scattered simple hairs, lower surface glabrescent with few simple or paired hairs on midrib and main nerves, lateral nerves 3–5 pairs, basal pair well developed (see Fig. 4/3 & 4/4, p. 10). Inflorescence terminal, rhachis and branchlets with long brown setulose hairs. Sepals oblong-lanceolate, 4–5 mm. long, 1–1·5 mm. wide, fringed with club-shaped glandular hairs. Petals white to yellowish, narrowly oblong-lanceolate, 4·5–5·5 mm. long, 1–1·5 mm. wide. Stamens 10 ; long-stamened flowers, 5 episepalous stamens to 5 mm. long, 5 epipetalous to 3 mm. long ; short-stamened flowers, 5 episepalous stamens to 3·5 mm. long, 5 epipetalous to 1·5 mm. long, filament-tube ± 1 mm. long, anthers not apiculate. Ovary ovoid ± 1 mm. long, densely pilose, styles of long-stamened flowers < 1 mm. long, recurved, styles of short-stamened flowers to 5 mm. long, pilose. Fruit an obliquely ovoid follicle, to 1·7 cm. long, 1 cm. diam. ; pericarp red, with velutinous indumentum. Seed shining black, to 1 cm. long, 7 mm. diam., slightly flattened ; aril to 2·5 mm. long, oblique, lobulate.

KENYA. Tana River District : Bura, 19 Oct. 1945 (fr.), *Mrs. Joy Adamson* 164 *in Bally* 6064 !
TANGANYIKA. Pangani District : Makinjumbi, *Scheffler* 247 !
ZANZIBAR. Wanda country, Kurekwe, 3 Dec. 1930 (fr.) *Greenway* 2636 ! ; Kombeni cave-wells, 12 Oct. 1930 (fl.), *Vaughan* 1626 !
DISTR. **K7** ; **T3** ; **Z** ; apparently restricted to this area
HAB. Riverine forest, coastal forest and woodland, 0–150 m.

NOTE. Records for this species are very few; a sterile specimen from the *Afzelia-Trachylobium* coastal forest at Buda Mafisini, about 13 km. WSW. of Gazi, *Drummond & Hemsley* 3801, suggests that the plant may be found in other east coast forests and it may be typically a liane of such forests.

5. SANTALOÏDES

Schellenb., Beitr. Anat. Syst. Conn. 76 (1910), and in E.P. IV. 127 : 119 (1938), *nom. conserv. prop.*

Shrubs, scandent or otherwise or lianes. Leaves imparipinnate, rarely simple, leaflets entire. Inflorescence an axillary or subterminal panicle. Flowers pentamerous, androecium and gynoecium heteromorphic. Sepals strongly imbricate, cartilaginous, margin usually ciliate. Petals longer than sepals, glabrous. Stamens 10 ; filaments glabrous, connate into short tube at base. Carpels glabrous or slightly hairy, only one matures ; stigma subcapitate ; ovules basally inserted. Fruit an oblong to ovoid follicle, slightly arcuate, calyx persistent and clasping base ; dehiscence by ventral suture or by series of irregular basal splits ; pericarp thin, tough, longitudinally striate with anastomosing fibres. Seed totally or partially enclosed by, but separate from, a thick fleshy aril ; cotyledons plano-convex, fleshy, radicle apical, endosperm absent.

The genus is mainly tropical Asian, but it occurs in Madagascar and has a few species in tropical and subtropical Africa.

S. splendidum (*Gilg*) [*Schellenb. ex*] *Engl.* in V.E. 3 (1) : 327 (1915) ; E.P. IV. 127 : 140 (1938) ; T.T.C.L. 169 (1949) ; F.C.B. 3 : 82, t. 5 (1952).

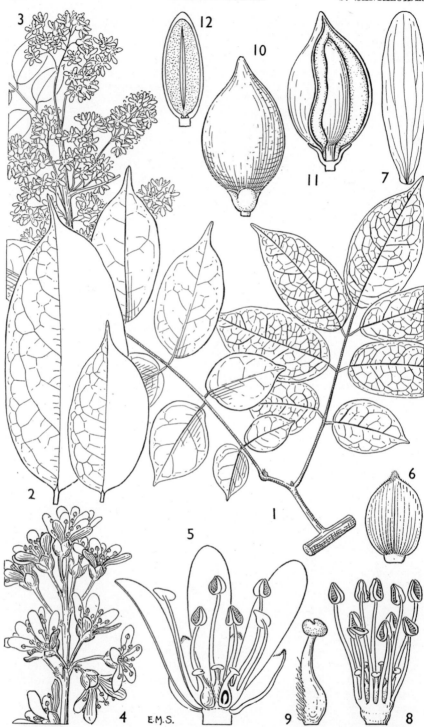

FIG. 5. *SANTALOÏDES SPLENDIDUM*—1, leafy branchlet, × ⅔; 2, leaflets to show size variation, × ⅔; 3, inflorescence, × ⅔; 4, part of inflorescence to show flower arrangement, × 2; 5, long-stamened flower, l.s., × 8; 6, sepal, × 8; 7, petal, × 8; 8, long-stamened flower with sepals and petals removed, × 8; 9, single carpel of long-stamened flower, × 16; 10, fruit, × 2; 11, fruit with part of pericarp and aril cut away to show seed, × 2; 12, seed l.s., × 2. 1, 2, 10–12, from *Drummond & Hemsley* 3427; 3–9 from *Eggeling* 5548.

Type : Belgian Congo, Lusambo Province, Mukenge, *Pogge* 744 (B, holo. †)

Shrub, sometimes scandent, or liane, height to 30 m. ; young branches glabrescent, lenticels inconspicuous. Leaf-rhachis to 20 cm. long, glabrescent. Leaflets 2–5 pairs, ovate to ovate-elliptic, 7–10 cm. long, 3·5–6 cm. wide, apex usually long, narrowly acuminate or shorter and more bluntly acuminate, base symmetrical or oblique, rounded to broadly cuneate, chartaceous ; glabrous, upper surface shining green, lower surface dull with raised reticulate venation. Inflorescence axillary, fascicled ; rhachis to 6 cm. long, glabrescent ; branchlets slender. Flowers fragrant. Sepals broadly ovate, 3–4 mm. long, 3–3·5 mm. wide, margin ciliate, otherwise glabrous. Petals white or pinkish, oblong-lanceolate, to 8 mm. long, 2·5 mm. wide. Stamens 10 ; long-stamened flowers, 5 episepalous stamens to 5·5 mm. long, 5 epipetalous to 4 mm. long; short-stamened flowers, 5 episepalous stamens to 4 mm. long, 5 epipetalous to 2·5 mm. long; filaments slightly dilated at base, anthers obcordate. Ovary subglobose, ± 1 mm. long, pilose ; styles of long-stamened flowers 1–1·5 mm. long, short-stamened flowers to 5 mm. long, terete, glabrescent. Fruit (said to be edible with a cherry flavour) an obliquely ovoid follicle to 2 cm. long, 1·2 cm. diam., apex acute and often somewhat bulbous ; dehiscence by series of irregular basal slits ; pericarp red when mature. Seed narrowly ovoid, compressed, to 1·3 cm. long, 7 mm. diam. ; totally enclosed within a pale juicy aril. Fig. 5.

UGANDA. West Nile District : Koboko, Mar. 1935 (fr.), *Eggeling* 1835 ! and West Madi, Amua Valley, Dec. 1944 (fl.), *Eggeling* 5548 !
KENYA. Kwale District : Shimba Hills, Mwele Mdogo Forest, 28 Aug. 1953 (fallen fr.), *Drummond & Hemsley* 4034A !
TANGANYIKA. Lushoto District : E. Usambaras, Nderema–Monga road, 23 July 1953 (fr.) *Drummond & Hemsley* 3427 !
DISTR. **U1, K7, T3** ; widespread in tropical Africa from French Guinea eastwards to the A.-E. Sudan, and southwards to the Western Province, Northern Rhodesia

SYN. *Rourea splendida* Gilg in E.J. 14 : 321 (1891)
     R. gudjuana Gilg in E.J. 14 : 323 (1891). Type : border A.-E. Sudan and French Equatorial Africa, Dar Fertit, Dem Gudju, *Schweinfurth*, Series 3, 223 (B, holo. †, K, iso. !)
     Santaloïdes gudjuanum (Gilg) Schellenb. in E.J. 55 : 454 (1919) ; E.P. IV. 127 : 138, fig. 24 (1938)

HAB. Woodland, riverine forest, and lowland rain-forest, 0–1000 m.

NOTE. The Kenya record is based on a gathering of fallen fruits only. Birds and monkeys seem to be very fond of the fruits and often indicate the presence of the liane in the forest canopy by the strewn debris of old pericarps and partially eaten remnants which can be found in abundance under the fruiting lianes.

6. BYRSOCARPUS

Schumach. & Thonn., Beskr. Guin. Pl. 226 (1827) ; Schellenb. in E.P. IV. 127 : 146 (1938)

Erect or scandent shrubs, small trees or lianes, usually with prominent lenticels. Leaves imparipinnate, leaflets small and numerous or larger and less numerous, opposite or sub-opposite ; lateral nerves fine, running ± straight towards margin. Inflorescences axillary, racemose, few-flowered. Flowers precocious, pentamerous, androecium and gynoecium dimorphic. Sepals strongly imbricate, clasping base of corolla, glabrescent or puberulous, margins ciliate. Petals longer than sepals, glabrous. Stamens 10, filaments connate into a short basal tube, glabrescent. Ovary hairy ; style terete ; stigma capitate or subcapitate, ovules inserted basally. Fruit a ±

ovoid follicle, sometimes curved, apex mucronate ; calyx persistent, closely
clasping the fruit base. Dehiscence by ventral suture, seed extruded and
held by pericarp. Seed ovoid, totally or partially enclosed by brightly
coloured aril ; testa and aril fused to form a compound fleshy pseudo-aril ;
cotyledons fleshy, radicle apical or laterally displaced, endosperm absent.

A genus mainly of tropical Africa ; two species known from Madagascar.

Leaflets 2–4 pairs ; apex of leaflet ± acuminate .   . 1. *B. boivinianus*
Leaflets 5–14 pairs ; apex of leaflet ± rounded or
  emarginate :
  Leaflets 5–8 pairs ; terminal leaflet narrowly cuneate
    at base, lateral leaflets oblique with very unequal
    sides .   .   .   . 2. *B. coccineus*
  Leaflets 7–14 pairs ; terminal leaflet broadly cuneate
    to rounded at base, lateral leaflets not or only
    slightly oblique .   .   .   .   .   . 3. *B. orientalis*

1. **B. boivinianus** (*Baill.*) *Schellenb.*, Beitr. Anat. Syst. Conn. 40 (1910) ;
E.P. IV. 127 : 155 (1938) ; T.T.C.L. 167 (1949). Type : Kenya, Mombasa,
Boivin (P, holo.)

Shrub, scandent or otherwise, or small tree to 4 m. high. Young stems
reddish brown with numerous small lenticels, older stems grey. Leaf-
rhachis 6–17 cm. long, glabrous. Leaflets 2–4 pairs, elliptic to ovate,
2·5–6 cm. long, 1·5–4 cm. wide, apex acuminate or sub-acuminate, base
rounded ; lower lateral leaflets ± ovate, sub-oblique, with rounded to ±
cordate base ; chartaceous to sub-coriaceous, glabrous and ± glaucous when
dry. Inflorescence rhachis to 7 cm. long, glabrous. Sepals ovate to broadly
ovate, to 3·5 mm. long, 3 mm. wide, glabrescent. Petals ligulate or narrowly to broadly elliptic, to 12 mm. long, 3 mm. wide. Stamens 10 ; long-
stamened flowers, 5 episepalous stamens, 7·5–11 mm. long, 5 epipetalous
5·5–9 mm. long ; short-stamened flowers, 5 episepalous stamens to 3·5 mm.
long, 5 epipetalous to 2·5 mm. long ; filament-tube 1–2·5 mm. long. Ovary
ovoid, to 1·5 mm. long, hirsute ; styles of long-stamened flowers to 3 mm.
long, styles of short-stamened flowers to 8·5 mm. long. Follicle yellow to
scarlet when mature, to 2 cm. long, 1 cm. diam. Seed to 1·6 cm. long, 7 mm.
diam., almost enclosed within bright red pseudo-aril ; exposed tip of testa
black ; radicle ventral.

KENYA. Kilifi District : Arabuko-Sokoke Forest, June 1937 (fr.), *Dale* 3775 !
TANGANYIKA. Handeni District : 30 km. S. of Handeni, 10 Mar. 1953 (fl.), *Drummond
  & Hemsley* 1467 ! ; Kilosa, Jan. 1926 (fr.), *B. D. Burtt* 74 ! ; Rufiji District : Mafia
  Island, Oct. 1937 (fl.), *Greenway* 5383 !
DISTR. K7 ; T3, 6, 8 ; Portuguese East Africa
HAB. Coastal forest and bushland but extending up to 200 km. inland along the main
  river systems in forest and woodland complexes, 0–750 m.

SYN. *Rourea boiviniana* Baill. in Adansonia 7 : 231 (1867)
      Byrsocarpus ovatifolius Baker in F.T.A. 1 : 452 (1868). Type : southern
        Tanganyika border, Rovuma River, *Meller* (K, holo. !)
      B. maximus Baker in F.T.A. 1 : 453 (1868) ; Schellenb. in V.E. 3 (1) : 324,
        fig. 211 (1915). Type : southern Tanganyika border, Rovuma River, *Kirk*
        (K, holo. !)
      Rourea maxima (Baker) Gilg in P.O.A. C : 192 (1895)
      R. ovatifolia (Baker) Gilg in P.O.A. C : 192 (1895)
      R. usaramensis Gilg in P.O.A. C : 192 (1895). Types : Tanganyika, [prob.
        Bagamoyo or Uzaramo Districts] Dunda, *Stuhlmann* 6420 ; Dilangilo,
        Stuhlmann 6641; Kikuli, *Stuhlmann* 6780 ; and Maguli, *Stuhlmann* 7091 (all
        B, syn. †)
      R. goetzei Gilg in E.J. 28 : 394 (1900). Type : Tanganyika, Kilosa District, E.
        slopes of Vidunda Mts., bank of Ruaha River, *Goetze* 415 (B, holo. †, K, iso. !,
        written up as *R. goetzeana* Gilg)

Byrsocarpus usaramensis (Gilg) Schellenb., Beitr. Anat. Syst. Conn. 43 (1910)
B. goetzei (Gilg) Greenway in T.T.C.L., part 1, 41 (1940)

VARIATION. Specimens from areas with markedly seasonal water supply have a thick rough bark with very dense pale raised lenticels giving a coral like appearance to the surface. The leaflets of plants from such areas are broader, more coriaceous and very shortly acuminate. This species in common with others of the genus sometimes flowers on young vigorous shoots arising from burnt-back stools and in this case the inflorescences are long and lax with very few, large flowers, the sepals and petals tending to be longer and broader than the normal dimensions for the species.

2. **B. coccineus** *Schumach. & Thonn.*, Beskr. Guin. Pl., 226 (1827) ; F.T.A. 1 : 452 (1868) ; E.P. IV. 127 : 148 (1938) ; T.T.C.L. 167 (1949) ; F.C.B. 3 : 91 (1952). Type : Guinea, *Thonning* (C, holo.)

Straggling or scandent shrub, height to 3 m. Branches reddish brown with numerous lenticels. Leaf-rhachis 4–12 cm. long, pilose when young, later glabrescent. Leaflets 4–9 (usually 5–6) pairs, oval, elliptic to oblong, 0·5–3 cm. long, 0·3–2 cm. wide, apex rounded to emarginate, base ± rounded to cuneate, lateral leaflets oblique ; lower surface of young leaflet pilose especially on midrib, later glabrescent. Inflorescence rhachis to 4 cm. long, slender, ± glabrous. Flowers fragrant. Sepals broadly ovate, to 2 mm. long, 1·9 mm. wide, margins ciliate. Petals white to pale yellow, narrowly elliptic, to 9 mm. long, 2 mm. wide. Stamens 10 ; long-stamened flowers, 5 episepalous stamens to 6 mm. long, 5 epipetalous to 4·5 mm. long ; short-stamened flowers, 5 episepalous stamens to 2·5 mm. long, 5 epipetalous to 2 mm. long ; filament-tube very short. Ovary ± 1 mm. long, styles of long-stamened flowers to 1·5 mm. long, styles of short-stamened flowers to 6 mm. long. Mature follicle red, to 1·8 cm. long, 1 cm. diam., sepals clasping base at first, later reflexing. Seed 1·5 cm. long, 8 mm. diam., pseudo-aril yellow or orange, lobed, enclosing seed almost to apex ; exposed testa black ; radicle ventral.

TANGANYIKA. Kigoma District : Malagarasi, *Peter* 35855, 36074 & 36169
DISTR. T4 ; widely distributed to the west from French Guinea to Angola, Belgian Congo and southwards to Northern Rhodesia, apparently just reaching our area on the eastern shores of Lake Tanganyika
HAB. Open areas in woodland and on edges of forest, 0–1500 m.

SYN. *Rourea inodora* De Wild. & Th. Dur. in Ann. Mus. Congo, Bot. sér. I, 1 : 71, t. 36 (1899). Type : Belgian Congo, Mayombe, Kembo, *Dewèvre* 442 (BR, holo.)
     Byrsocarpus inodorus (De Wild. & Th. Dur.) Schellenb. in V.E. 3 (1) : 325 (1915) ; T.T.C.L., part 1 : 41 (1940)
     B. puberulus Schellenb. in E.P. IV. 127 : 150 (1938). Types : Belgian Congo, Kinanga–Kisantu, *Oddon in Gillet* 1876 (BR, holo.)

3. **B. orientalis** *(Baill.) Baker* in F.T.A. 1 : 452 (1868) ; E.P. IV. 127 : 151 (1938) ; T.T.C.L. 167 (1948). Type : Kenya, Mombasa, *Boivin* (P, holo.)

Shrub, sometimes scandent, or small tree, height to 6 m. ; branches reddish brown or grey-brown with conspicuous lenticels, young twigs sparsely hairy to densely pubescent. Leaf-rhachis to 25 cm. long, subglabrous or pubescent. Leaflets 6–14 pairs, elliptic to elliptic-oblong, 1·2–4 cm. long, 0·7–1·9 cm. wide, rounded and usually mucronate at apex, rounded to broadly cuneate at base ; upper surface glabrous, lower surface glabrous or with scattered hairs, midrib sometimes densely hairy. Inflorescence rhachis to 5 cm. long, glabrous or pubescent. Flowers fragrant. Sepals ovate, to 3 mm. long, 2 mm. wide. Petals white to yellow, ligulate or narrowly elliptic, 0·7–1·1 cm. long, 1·5–3·5 mm. wide. Stamens 10 ; long-stamened flowers, 5 episepalous stamens to 7 mm. long, 5 epipetalous to 4 mm. long ; short-stamened flowers, 5 episepalous stamens to 3 mm. long,

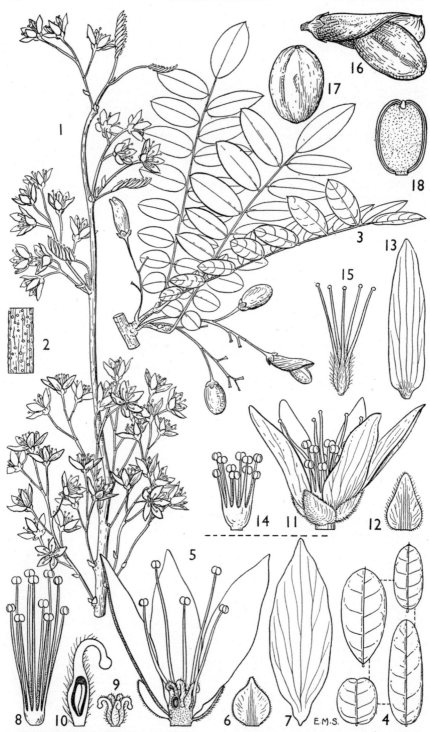

FIG. 6. *BYRSOCARPUS ORIENTALIS*—**1,** flowering branch with young leaves, × 1 ; **2,** portion of bark showing lenticels, × 2 ; **3,** mature leaves and fruits, × ⅔ ; **4,** leaflets to show variation in shape and size, × ⅔ ; **5,** long-stamened flower, l.s., × 6 ; **6,** sepal, × 6 ; **7,** petal, × 6 ; **8,** androecium, × 6 ; **9,** gynoecium, × 6 ; **10,** carpel cut open to show ovules, × 15 ; **11,** short-stamened flower, × 6 ; **12,** sepal, × 6 ; **13,** petal, × 6 ; **14,** androecium, × 6 ; **15,** gynoecium, × 6 ; **16,** mature fruit with extruded seed, × 1½ ; **17,** seed, × 1½ ; **18,** seed l.s., × 1½. 1, 2, 5–10, from *Bruce* 271 ; 3, 16–18, from *Drummond & Hemsley* 1132 ; 4, from various specimens ; 11–15, from *Jeffery* 166.

5 epipetalous to 2 mm. long ; filament-tube 1–2 mm. long. Ovary ovoid, ± 1 mm. long, densely pilose ; styles of long-stamened flowers to 1·5 mm. long, usually recurved, styles of short-stamened flowers to 4·5 mm. long. Mature follicle reddish, to 2 cm. long, 1 cm. diam., calyx spreading at maturity. Seed to 1·5 cm. long, 8 mm. diam., totally enclosed within bright red pseudo-aril; radicle apical.   Fig. 6.

KENYA. Kilifi District : Sokoke, 17 Apr. 1945 (fl.), *Jeffery* K166 !
TANGANYIKA. Lushoto District : E. Usambara Mts., Mgambo, 26 Mar. 1940 (fr.), *Greenway* 5930 ! ; Pangani District : S. bank of Pangani river, between Hale & Makinjumbi, 1 July 1953 (fr.), *Drummond & Hemsley* 3133 ! ; Iringa District : near Malangali, Nov. 1928 (fl.) *Haarer* 1644 !
ZANZIBAR. Near Sebeleni, 22 Apr. 1933 (fl.), *Vaughan* 2105 !
DISTR. K7 ; T3, 4, 6–8 ; Z ; and southwards to Portuguese East Africa, Nyasaland, the Rhodesias, SE. Belgian Congo, Angola and Bechuanaland
HAB. Forest margins, woodland and bushland complexes, 0–2000 m.

SYN. *Rourea orientalis* Baill. in Adansonia 7 : 230 (1866/67)
   R. ovalifoliolata Gilg in E.J. 14 : 327 (1891). Type : Kenya, Mombasa, Mar. 1876 (fl.), *Hildebrandt* 1934 (B, holo. †, K, iso. !)
   R. macrantha Gilg in E.J. 27 : 393 (1900). Type : Tanganyika, Kilosa District : Uhehe, by Ruaha River, Jan. 1898, *Goetze* 417 (B, holo. †, K, iso. !)
   [*Byrsocarpus baillonianus* Gilg ex Schellenb., Beitr. Anat. Syst. Conn. 40 (1910), nom. nud. based on Gilg's MS. name in the Berlin Herbarium.]
   B. ovalifoliolatus (Gilg) Schellenb., Beitr. Anat. Syst. Conn. 42 (1910)
   B. tomentosus Schellenb. in E.J. 55 : 452 (1919). Type : Tanganyika, Songea, *Busse* 804 (B, holo. †, EA, iso. !)
   B. usambaricus Schellenb. in E.P. IV. 127 : 154 (1938). Type : Tanganyika, Lushoto District, Sigi Valley, Longusa, *Peter* 40014 (B, holo. †)

VARIATION. A very variable species, the type of which was described from the extreme NE. corner of the distributional range. Presence or absence of indumentum, number, shape and texture of leaflets show a range of variation in which it is impossible to define any clear-cut segregation. In general, specimens from the northern and east coastal areas tend to possess subglabrous young stems and undersurfaces of leaflets, the density of indumentum increasing in the south-western districts and becoming a conspicuous feature in the Rhodesian material. This tendency seems to be greatest in specimens from habitats subject to annual burning. Leaflet-number is a further character showing a geographical correlation, the lower number tending to occur in the north-east and increasing towards the higher figure in the south-west and Rhodesia. Leaflet-shape and texture appear to be habitat variation on a very local scale, depending upon incidence of water supply, temperature and light factors, Fig. 6/4 illustrates the extremes of shape shown in herbarium specimens.

7. **JAUNDEA**

Gilg in N.B.G.B. 1 : 66 (1895) ; Schellenb. in E.P. IV. 127 : 161 (1938)

Byrsocarpus subgen. *Jaundea* Schellenb., Beitr. Anat. Syst. Conn. 43 (1910), *ex parte*

Shrubs, sometimes scandent, small trees or lianes. Leaves imparipinnate, opposite or sub-opposite. Inflorescence terminal or axillary, paniculate. Flowers pentamerous, androecium and gynoecium heteromorphic. Sepals imbricate, puberulous externally. Petals longer than sepals, glabrous. Stamens 10 ; filaments connate into a short tube at base, glabrous. Carpels pilose, style glabrous or pilose ; stigma capitate ; ovules inserted basally. Fruit a curved, ovoid or subcylindrical follicle, apex rounded, calyx persisting and clasping base, glabrous. Dehiscence by ventral suture. Seed ovoid ; testa fused with aril to form a compound fleshy pseudo-aril ± enclosing the seed ; cotyledons fleshy ; radicle displaced laterally in a ventral position ; endosperm nil.

The genus is confined to tropical Africa ; very closely allied to *Byrsocarpus* but differing mainly in the following characters : flowering when in full foliage, in contrast

FIG. 7. *JAUNDEA PINNATA*—**1**, flowering branch with leaves, × ⅔ ; **2**, flower, × 4 ; **3**, long-stamened flower with some sepals and petals removed, × 4 ; **4**, calyx opened out to show inner surface of sepals, × 4 ; **5**, sepal to show exterior surface, × 8 ; **6**, petal, × 8 ; **7**, stamens, × 16 ; **8**, gynoecium, × 8 ; **9**, single carpel, × 8 ; **10**, mature fruit with extruded seed, × 1 ; **11**, seed, × 1. 1–9, from *Wallace* 452 ; 10 and 11, from *Drummond & Hemsley* 4375.

to the precocious flowering of *Byrsocarpus* species ; paniculate inflorescence ; well marked arcuate ascending lateral nerves of the leaflets and the possession of a con-spicuous ventral groove on the seed running to the base of the pseudo-aril.

J. pinnata (*Beauv.*) *Schellenb.* in Candollea 2 : 92 (1925) ; E.P. IV. 127 : 164, fig. 29 (1938) ; F.C.B. 3 : 87, Pl. VI. (1952). Type : Nigeria, Awarri [Oware], *Beauvois* (G, holo.)

Shrub, small tree or liane, height to 25 m. Young branches puberulous, soon becoming glabrous, older branches with brownish red bark and pale lenticels. Leaf-rhachis 6–18 cm. long, glabrous. Leaflets 2–4 pairs, elliptic to elliptic-oblong, 2·5–16 cm. long, 1·5–7 cm. wide, apex shortly apiculate, base rounded to cuneate, chartaceous to sub-coriaceous ; midrib above deeply impressed, both surfaces glabrous, lateral nerves 4–6 pairs, arcuate, ascend-ing. Inflorescence axillary or sub-terminal, fascicled, rhachis to 9 cm. long, pilose to densely brown pubescent. Flowers fragrant. Sepals ovate, to 3 mm. long, 2 mm. wide, margin ciliate. Petals white to creamy yellow, narrowly elliptic to ligulate, to 10 mm. long, 2·5 mm. wide. Stamens 10 ; long-stamened flowers, 5 episepalous stamens to 3·5 mm. long, 5 epipetalous to 2·5 mm. long ; short-stamened flowers, 5 episepalous stamens to 1·5 mm. long, 5 epipetalous to 1 mm. long, filaments flattened near base. Ovary sub-globose, to 1 mm. long, densely pilose ; styles of long-stamened flowers to 1·5 mm. long, short-stamened flowers to 4 mm. long, glabrous : stigma papillose. Fruit to 2·8 cm. long, 1·7 cm. diam., pericarp reddish or greenish-yellow at maturity, tough and leathery. Seed to 2·1 cm. long, 1·4 cm. diam., pseudo-aril red, almost enclosing seed, with ventral groove running to base. Seed held by lips of pericarp after dehiscence. Fig. 7.

UGANDA. Kigezi District : Kanungu, June 1939 (fl. & fr.), *Purseglove* 802 ! ; Mengo District : Kajansi Forest, mile 10, Entebbe road, Oct. 1938, *Chandler* 2476 !
KENYA. Kiambu District : Uplands, Limuru, 15 Oct. 1950 (fr.), *Verdcourt* 359 ! ; Teita Hills, Chawia Forest, Bura bluff, 17 Sept. 1953 (fr.), *Drummond & Hemsley* 4375 !
TANGANYIKA. Lushoto District : W. Usambara Mts., Shagai Forest, 2 km. SE. Sunga, 2 Mar. 1953 (fl.), *Drummond & Hemsley* 1407 ! and 17 May 1953 (fr.), *Drummond & Hemsley* 2585 ! ; Morogoro District : Uluguru Mts., 18 Nov. 1932 (fl.), *Wallace* 452 ! and 21 Nov. 1932 (fl.), *Wallace* 462 !
DISTR. U2, 4 ; K4, 7 ; T1–3, 6, 7 ; extends from French Guinea in the west, to the A.-E. Sudan in the north and southwards through Northern Rhodesia to Angola
HAB. Typically a liane of lowland and upland rain-forest, but occurs as a shrub or small tree in forest remnants and cleared agricultural areas, 0–2500 m.

SYN. *Cnestis pinnata* Beauv., Fl. Oware 1 : 98, t. 60 (1804)
  Rourea monticola Gilg in N.B.G.B. 1 : 68 (1895). Types : Tanganyika, Uluguru Mts., Nglewenu, *Stuhlmann* 8957 (B, syn. †) and central Uluguru Mts., Kifuru, *Stuhlmann* 9071 (B, syn. †)
  R. albido-flavescens Gilg in E.J. 30 : 316 (1901). Type : Tanganyika, Kinga Mts., Manganyema, *Goetze* 1212 (B, holo. †)
  Byrsocarpus monticolus (Gilg) Schellenb., Beitr. Anat. Syst. Conn., 44 (1910)
  Jaundea monticola (Gilg) Schellenb. in E.J. 55 : 461 (1919) ; E.P. IV. 127 : 166 (1938) ; F.P.N.A. 1 : 258, Pl. XXIV (1948) ; T.T.C.L. 168 (1949) ; F.C.B. 3 : 88 (1952)
  [*Rourea sp.* sensu auct. in Check List Uganda Trees and Shrubs 39 (1935) based on *Mahon* s.n. !]
  Byrsocarpus albo-flavescens (Gilg) Greenway in T.T.C.L., part 1, 41 (1940), in error for *albido-flavescens*

VARIATION. This widespread species shows considerable variation in the degree of reticulation and prominence of venation of the leaflet lower surface. Specimens from the higher altitudes of Kenya possess a close and prominently raised reticulate vena-tion which contrasts with the more typical lower altitude plants from Uganda, E. Usambara and Uluguru Mts. A further character-variation which seems to be cor-related with altitude, is the density of indumentum on the inflorescence rhachis ; specimens from the upland rain forests are sub-glabrous but those from the lower altitudes show varying degrees of hairiness and merge with the typical facies of the species as shown by the Congo and W. African material.

8. **ELLIPANTHUS**

Hook. f. in G.P. 1 : 434 (1862) ; E.P. IV. 127 : 181 (1938)

Shrubs or small trees, with unifoliolate, alternate leaves. Inflorescences racemose or paniculate, axillary. Flowers with dimorphic androecium and gynoecium. Sepals 5, ± valvate. Petals 5, glabrous to pubescent. Stamens 10; 5 episepalous fertile, longer than 5 epipetalous staminodes ; all connate into a basal tube, tube hairy internally. Carpel solitary, ovary and style villous, stigma capitate, ovules inserted on ventral suture. Fruit an ovoid, obliquely pyriform or cymbiform follicle, acute at apex, narrowed into a basal stipe, densely hairy externally ; dehiscence by ventral suture, seed retained but exposed within pericarp, the latter splitting from apex towards base, pericarp dry and tough at maturity. Testa of seed dark and shining, aril ± basal, oblique, membranous ; endosperm present, forming a thin layer around cotyledons, the latter fleshy, somewhat flattened, radicle apical.

A genus otherwise confined to tropical Asia.

E. hemandradenioïdes *Brenan* in Hook., Ic. Pl. 35 : t. 3452 (1947). Type : Kenya, Malindi District, Mida, *Dale* 3876 (K, holo. !, EA, iso. !)

Small tree to 10 m. high, branchlets slender, puberulous with appressed hairs when young, becoming glabrous ; bark purplish brown, longitudinally striate, with scattered lenticels. Lamina narrowly to broadly ovate (4·5–) 6·5–11·5 cm. long, (2–) 2·5–5·2 cm. wide, apex narrowly acuminate, base broadly cuneate to rounded, coriaceous, upper surface with scattered appressed hairs, becoming glabrous with age, lower surface hairy, especially on midrib and nerves ; leaf margin thickened, lateral nerves 4–6 pairs, arcuate, ascending. Inflorescence a few-flowered panicle, with dense grey-brown indumentum ; rhachis 0·4–2 cm. long. Sepals ovate, ± 2 mm. long, 1·2 mm. wide, pubescent. Petals white, straplike, 4·5–7·5 mm. long, 1–1·5 mm. wide, puberulous. Filaments of fertile stamens to 3 mm. long, subulate, anthers ellipsoid, ± 1·5 mm. long ; staminodes to 2·2 mm. long, subulate, tapering to terminal papilla ; filament-tube to 1·5 mm. long. Ovary villous with long white hairs, to 2 mm. long and 1·5 mm. wide, style ± 3 mm. long, ovules attached near base of ventral suture. Fruit cymbi-form, up to 2·7 cm. long and 1·6 cm. wide, apex with short rostrum, basal stipe 3–7 mm. long, indumentum velutinous, cinerous. Seed black, oblong-ellipsoid, narrower at apex than base, slightly flattened, to 1·7 cm. long, 9 mm. diam. ; aril white, basally attached, ± 5 mm. long, slightly lobed, membranous. Dehiscence extends from apex to ⅔ length of follicle, seed slightly extruded, aril soon withers and disappears. Fig. 8.

KENYA. Kwale District : Buda Mafisini Forest, 13 km. WSW. of Gazi, 22 Aug. 1953 (fr.), *Drummond & Hemsley* 3954 ! ; Kilifi District : Mida, Oct. 1936 (fl.), *Dale* 3573 !, and Apr. 1938 (fr.), *Dale* 3876 !
DISTR. **K7** ; not known elsewhere
HAB. Understorey tree of *Afzelia-Trachylobium* lowland rain-forest on deep white sands, 0–100 m.

SYN. [*Hemandradenia sp.* sensu Dale, Woody Veg. Coast Prov. Kenya 24 (1939)]

9. **CONNARUS**

L., Gen. Pl., ed. 5, 305 (1754) ; Schellenb. in E.P. IV. 127 : 216 (1938)

Shrubs, trees or lianes, branches glabrous or with simple or sympodially branched hairs. Leaves imparipinnate, trifoliolate or rarely unifoliolate,

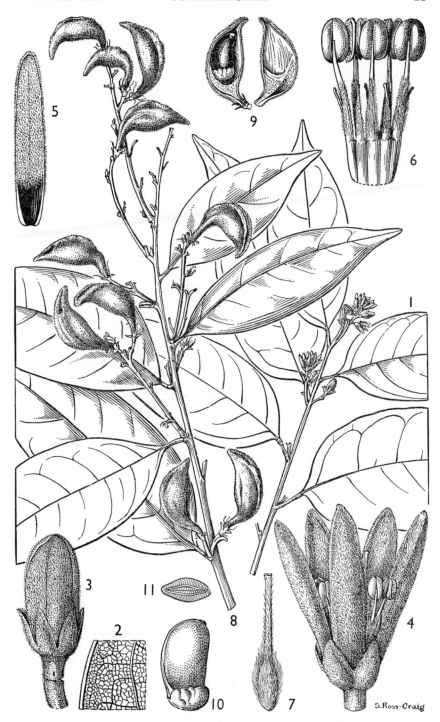

FIG. 8. *ELLIPANTHUS HEMANDRADENIOIDES*—**1,** flowering branch, × 1 ; **2,** leaf, upper surface, × 3 ; **3,** flower-bud, × 8 ; **4,** flower, × 8 ; **5,** petal, inner surface, × 8 ; **6,** androecium, × 12 ; **7,** carpel, × 8 ; **8,** fruiting branch, × 1 ; **9,** fruit opened longitudinally, showing seed, × 1 ; **10,** seed, × 2 ; **11,** seed, t.s., × 2. Note.—Drawings 10 and 11 somewhat reconstructed. 1–7, from *Dale* 3573 ; 8–11, from *Dale* 3876. Reproduced by permission of the Bentham-Moxon Trustees.

entire, opposite or sub-opposite. Inflorescence terminal, paniculate, or rarely axillary and cymose or racemose. Flowers with heteromorphic andro-ecium and gynoecium. Sepals 5, rarely 4, imbricate or rarely subvalvate. Petals 5, usually longer than sepals, glabrous to densely pubescent, some-times with glandular hairs ; both sepals and petals dotted with blackish glandular cavities. Stamens 10 ; epipetalous stamens sometimes staminodal or reduced with functionless anthers; filaments connate into short basal tube, anthers ± oblong, both filaments and anthers with scattered glandular papillae. Carpel solitary ; ovary ovoid, pubescent, style often with glandu-lar papillae, stigma expanded ; ovules inserted on ventral suture. Fruit an obliquely pyriform, fusiform or cymbiform follicle, apex obliquely mucronate or tapering into a curved rostrum, base narrowing into a slender stipe ; dehiscence usually along ventral suture ; pericarp woody, horny or tough and parchment-like, obliquely and conspicuously lined with raised striae, usually glabrous, sometimes with blackish glandular lacunae. Testa dark and shining, hilum lateral ; aril bilobed, spreading posteriorly to enclose seed-base ; cotyledons thick, radicle usually apical ; endosperm absent or little-developed.

A genus with pantropical distribution

C. longistipitatus *Gilg* in P.O.A. C : 191 (1895) ; Schellenb. in E.P. IV. 127 : 268, fig. 46 (1938), as *C. longestipitatus* ; T.T.C.L. 168 (1949) ; I.T.U., ed. 2, 100 (1952). Type : Tanganyika, Bukoba, *Stuhlmann* 3831 (B, holo. †)

Tree, shrub or liane to 30 m. high, young branches glabrescent with scattered lenticels ; old branches smooth, reddish brown. Leaf-rhachis to 18·5 cm. long, glabrous. Leaflets 2–4 pairs, usually 3, elliptic to elliptic-oblong, 6–16·5 cm. long, 3–6·5 cm. wide, apex shortly acuminate, base narrowed to petiolule, sub-coriaceous; both upper and lower surface ± glabrous ; lateral nerves 5–8 pairs, arcuate ascending. Inflorescence a large terminal panicle to 25 cm. long ; rhachis, branchlets and pedicels with dark brown pubescence. Sepals 5, ovate to elliptic, to 4 mm. long, 1·5 mm. wide, puberulous externally. Petals white, narrowly elliptic, to 8 mm. long, 2 mm. wide, slightly connate, puberulous along apical margin, otherwise glabrous. Stamens 10 ; long-stamened flowers, 5 episepalous stamens to 6 mm. long, 5 epipetalous to 3·5 mm. long ; short-stamened flowers, 5 episepalous stamens to 3 mm. long, 5 epipetalous to 1·5 mm. long ; filaments slightly flattened, with scattered capitate glands. Ovary sub-globose, ± 1 mm. long, puberulous; styles of long-stamened flowers to 2 mm. long, short-stamened flowers to 5 mm. long, terete with short hairs and capitate glands ; stigma capitate. Ovules affixed to ventral suture near base of ovary. Ripe follicle shortly and broadly cymbiform (see Fig. 9/14), to 2·8 cm. long, 1·8 cm. wide, apex obliquely apiculate, basal stipe 3–9 mm. long ; pericarp scarlet at maturity, tough and coriaceous, glabrous. Seed black, ovate-oblong, slightly flattened, 1·8 cm. long, 8 mm. diam. ; aril orange, fleshy, arising laterally and clasping base of seed. Fig. 9.

UGANDA. Ankole District : Kalinzu Forest near Kyanga camp, Dec. 1931 (fl. & fr.), *R. A. Gibson* 8/MA *in Brasnett* 387 ! ; Bunyoro District : Budongo Forest, Dec. 1934 (fl.), *Eggeling* 1556 *in F.H.* 1470 ! ; Mengo District : Kwewaga Forest near Entebbe, 4 Nov. 1950 (fl. & fr.), *Dawkins* 666 !
TANGANYIKA. Bukoba District : Kagera basin, Kiziba, 4 July 1947 (fr.), *Ford* 54 !
DISTR. **U**2, 4 ; **T**1 ; Belgian Congo
HAB. Rain-forest and seral stages in forest development, 1100–1500 m.

SYN. *C. stuhlmannianus* Gilg in P.O.A. C : 192 (1895). Type : Tanganyika, Bukoba, *Stuhlmann* 1128, 1576 (B, syn. †)

VARIATION. This species seems to vary greatly in life form. It flowers and fruits freely as a small shrub in the thickets on the raised ant-hills of the Kagera flats near Bukoba.

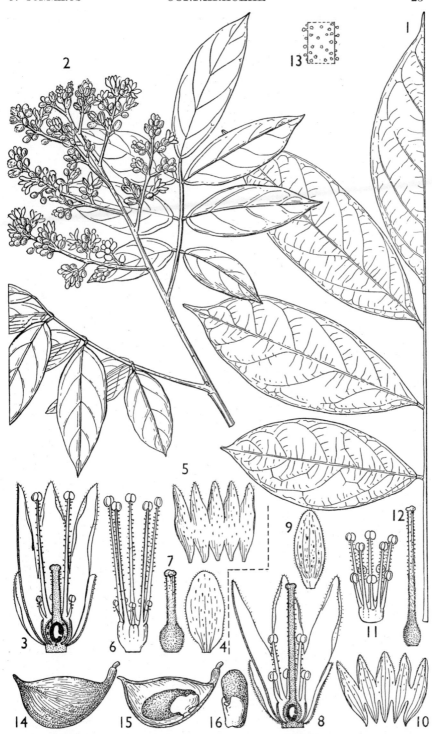

FIG. 9. *CONNARUS LONGISTIPITATUS*—**1**, large leaf, half only shown, × ½ ; **2**, flowering branch with young leaves, × ⅔ ; **3**, long-stamened flower, l.s., × 6 ; **4**, sepal, × 6 ; **5**, petals showing slight connation, × 3 ; **6**, androecium, × 6 ; **7**, carpel, × 6 ; **8**, short-stamened flower, l.s., × 6 ; **9**, sepal, × 6 ; **10**, petals showing slight connation, × 3 ; **11**, androecium, × 6 ; **12**, carpel, × 6 ; **13**, glandular hairs on style, × 30 ; **14**, fruit, × 1 ; **15**, fruit with half of pericarp removed to show seed, × 1 ; **16**, seed, × 1. 1, 2, 8–13, from *Eggeling* 1556 ; 3–7, from *Eggeling* 4411 ; 14–16, from *Dawkins* 666.

In the Mabira and the western Uganda forests it occurs as a small tree with a well marked trunk up to 25 cm. diam. ; typically it appears to be an upper canopy liane of the Uganda lake forests.

Imperfectly known species

C. sp.

Large forest liane to 30 m. or more high, stem base 15 cm. diam. Young stems with sparse hairs, soon becoming glabrous. Leaves imparipinnate ; rhachis to 11 cm. long, puberulous or glabrous, leaflets 2 pairs, elliptic, to 12 cm. long and 5 cm. wide, base rounded, apex acute to sub-acuminate ; upper surface with scattered medifixed hairs, becoming glabrous with age, lower surface glabrous with a reticulum of fine raised veins ; lateral nerves 5–7 pairs, arcuate ascending. Inflorescences terminal and sub-terminal, paniculate.

KENYA. Kwale District : Kwale, *Graham* Q290 *in F.D.* 1695 !
DISTR. **K7** ; known only from the above gathering
HAB. Lowland rain-forest, about 300 m.

NOTE. The cited specimen consists of one sheet in young bud. It would appear to be related to *C. longistipitatus* Gilg but differs in the appearance of the leaflet under-surface and the chartaceous texture of the leaflets. Further material is necessary to decide the identity of this plant.

INDEX TO CONNARACEAE